STUDENT LECTURE NOTEBOOK

Frederick K. Lutgens
Edward J. Tarbuck

Illustrated by Dennis Tasa

ESSENTIALS OF
GEOLOGY

TENTH EDITION

PEARSON

Prentice
Hall

Upper Saddle River, NJ 07458

Editor-in-Chief, Science: Nicole Folchetti
Publisher, Geosciences: Dan Kaveney
Assistant Editor: Sean Hale
Assistant Managing Editor, Science: Gina M. Cheselka
Project Manager, Science: Maureen Pancza
Supplement Cover Manager: Paul Gourhan
Supplement Cover Designer: Victoria Colotta
Operations Specialist: Amanda A. Smith
Senior Operations Supervisor: Alan Fischer
Cover Photo Credit: Jimmy Chin

© 2009 Pearson Education, Inc.
Pearson Prentice Hall
Pearson Education, Inc.
Upper Saddle River, NJ 07458

The author and publisher of this book have used their best efforts in preparing this book. These efforts include the development, research, and testing of the theories and programs to determine their effectiveness. The author and publisher make no warranty of any kind, expressed or implied, with regard to these programs or the documentation contained in this book. The author and publisher shall not be liable in any event for incidental or consequential damages in connection with, or arising out of, the furnishing, performance, or use of these programs.

Printed in the United States of America

10 9 8 7 6 5 4 3 2 1

ISBN-13: 978-0-13-604913-5
ISBN-10:   0-13-604913-3

Pearson Education Ltd., *London*
Pearson Education Australia Pty. Ltd., *Sydney*
Pearson Education Singapore, Pte. Ltd.
Pearson Education North Asia Ltd., *Hong Kong*
Pearson Education Canada, Inc., *Toronto*
Pearson Educación de Mexico, S.A. de C.V.
Pearson Education—Japan, *Tokyo*
Pearson Education Malaysia, Pte. Ltd.

Contents

To the Student...iv

CHAPTER 1   An Introduction to Geology...1-1

CHAPTER 2   Minerals: Building Blocks of Rocks... 2-1

CHAPTER 3   Igneous Rocks...3-1

CHAPTER 4   Volcanoes and Other Igneous Activity.....................................4-1

CHAPTER 5   Weathering and Soils...5-1

CHAPTER 6   Sedimentary Rocks..6-1

CHAPTER 7   Metamorphic Rocks...7-1

CHAPTER 8   Mass Wasting: The Work of Gravity...8-1

CHAPTER 9   Running Water...9-1

CHAPTER 10   Groundwater...10-1

CHAPTER 11   Glaciers and Glaciation...11-1

CHAPTER 12   Deserts and Wind...12-1

CHAPTER 13   Shorelines..13-1

CHAPTER 14   Earthquakes and Earth's Interior...14-1

CHAPTER 15   Plate Tectonics: A Scientific Theory Unfolds..........................15-1

CHAPTER 16   Origin and Evolution of the Ocean Floor.............................16-1

CHAPTER 17   Crustal Deformation and Mountain Building.........................17-1

CHAPTER 18   Geologic Time..18-1

CHAPTER 19   Earth's Evolution through Geologic Time................................19-1

Appendices..A-1

To the Student

This Student Lecture Notebook is designed to help you do your best in this geology course.

Key images from the textbook and every illustration from the Instructor's Transparency Set are reproduced in this notebook. Because you won't have to redraw the art in class, you can focus your attention on the lecture, annotate the art, and take your notes in this book.

Leave all your notes together or remove them for integration into a binder with other course materials.

NOTES:

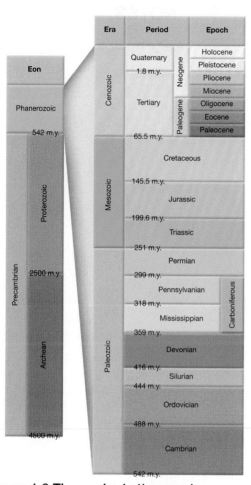

Figure 1.9 **The geologic time scale.**

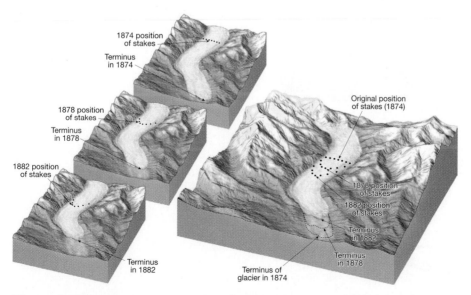

Figure 1.12 **Ice movement and changes in terminus at Rhône Glacier, Switzerland.**

Lutgens/Tarbuck, *Essentials of Geology*, 10e
© 2009 Pearson Prentice Hall, Inc.

NOTES:

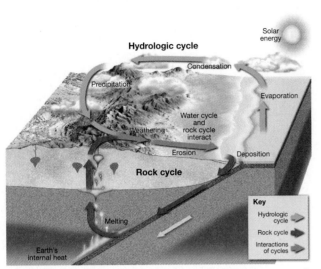

Figure 1.18 The Earth system involves many cycles.

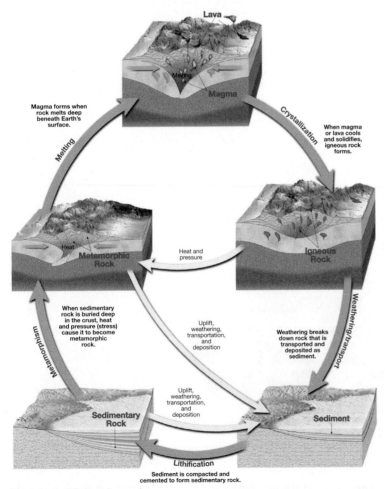

Figure 1.22 Rock cycle.

NOTES:

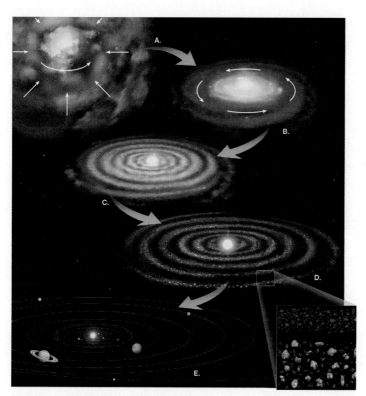

Figure 1.23 Nebular hypothesis.

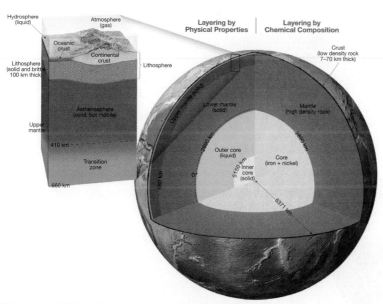

Figure 1.25 Earth's layered structure.

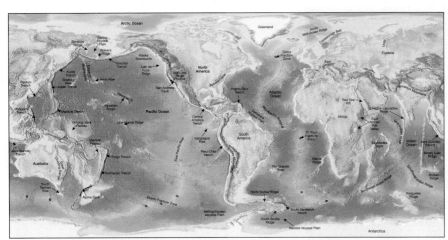

Figure 1.26 **Major surface features of the geosphere.**

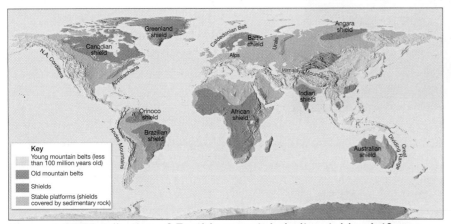

Figure 1.27 **Distribution of Earth's mountain belts, stable platforms, and shields.**

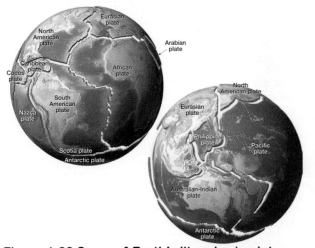

Figure 1.28 **Some of Earth's litospheric plates.**

NOTES:

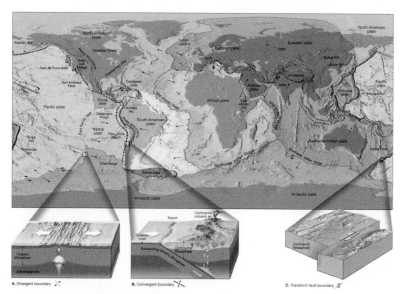

Figure 1.29 Earth's plates.

Figure 1.30 Convergent and divergent boundaries.

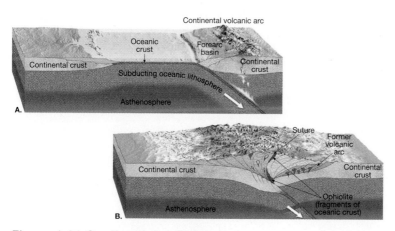

Figure 1.31 Continental collision.

NOTES:

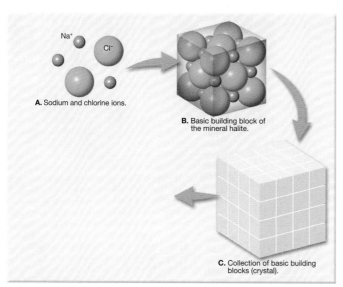

Figure 2.2A,B,C The arrangement of sodium and chloride atoms in the mineral halite.

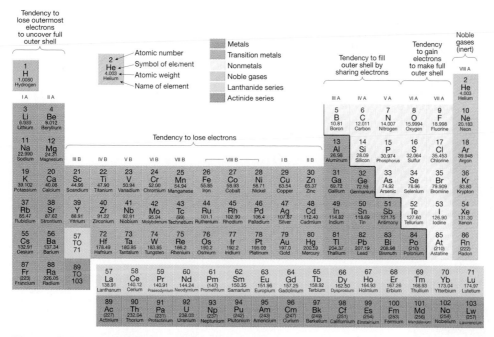

Figure 2.3 Periodic Table of the Elements.

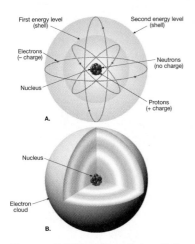

Figure 2.5 Two models of the atom.

Electron Dot Diagrams for Some Representative Elements							
I	II	III	IV	V	VI	VII	VIII
H•							He:
Li•	•Be•	•B•	•C•	•N•	:O•	:F•	:Ne:
Na•	•Mg•	•Al•	•Si•	•P•	:S•	:Cl•	:Ar:
K•	•Ca•	•Ga•	•Ge•	•As•	:Se•	:Br•	:Kr:

Figure 2.6 Dot diagrams for some representative elements.

A.

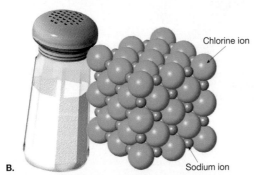

B.

Figure 2.7 Chemical bonding of sodium and chlorine atoms.

Lutgens/Tarbuck, *Essentials of Geology*, 10e
© 2009 Pearson Prentice Hall, Inc.

NOTES:

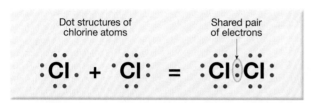

Figure 2.8 Dot diagrams illustrating the sharing of a pair of electrons in a chlorine molecule.

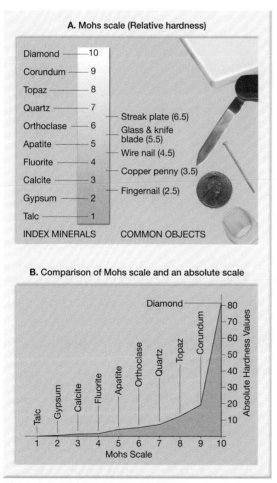

Figure 2.13 Hardness scales.

NOTES:

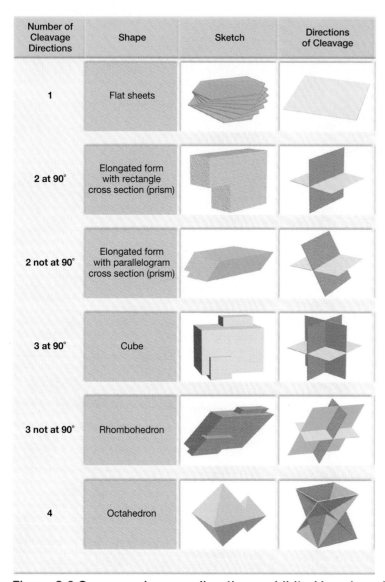

Number of Cleavage Directions	Shape	Sketch	Directions of Cleavage
1	Flat sheets		
2 at 90°	Elongated form with rectangle cross section (prism)		
2 not at 90°	Elongated form with parallelogram cross section (prism)		
3 at 90°	Cube		
3 not at 90°	Rhombohedron		
4	Octahedron		

Figure 2.6 **Common cleavage directions exhibited by minerals.**

NOTES:

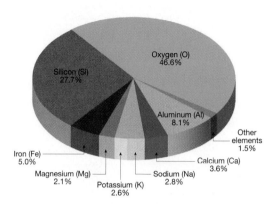

Figure 2.19 Relative abundance of the eight most abundant elements in continental crust.

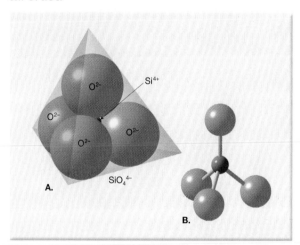

Figure 2.20 Two representations of the silicon-oxygen tetrahedron.

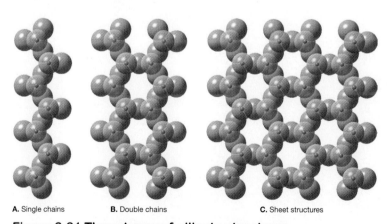

A. Single chains   **B.** Double chains   **C.** Sheet structures

Figure 2.21 Three types of silicate structures.

NOTES:

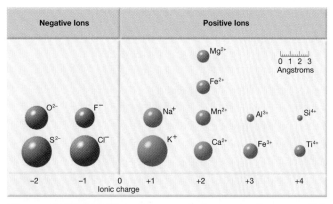

Figure 2.22 Relative sizes and charges of some of the most common elements in rock-forming minerals.

Mineral/Formula	Cleavage	Silicate Structure
Olivine group $(Mg, Fe)_2SiO_4$	None	Independent tetrahedron
Pyroxene group (Augite) $(Mg,Fe)SiO_3$	Two planes at right angles	Single chains
Amphibole group (Hornblende) $Ca_2(Fe,Mg)_5Si_8O_{22}(OH)_2$	Two planes at 60° and 120°	Double chains
Micas — Biotite $K(Mg,Fe)_3AlSi_3O_{10}(OH)_2$	One plane	Sheets
Micas — Muscovite $KAl_2(AlSi_3O_{10})(OH)_2$	One plane	Sheets
Feldspars — Potassium feldspar (Orthoclase) $KAlSi_3O_8$	Two planes at 90°	Three-dimensional networks
Feldspars — Plagioclase feldspar $(Ca,Na)AlSi_3O_8$	Two planes at 90°	Three-dimensional networks
Quartz SiO_2	None	Three-dimensional networks

Figure 2.23 Common silicate minerals.

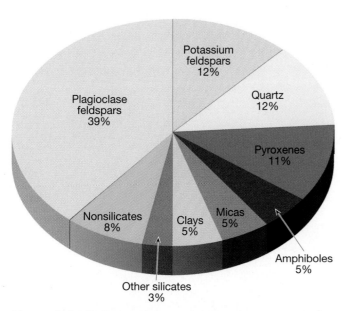

Figure 2.24 **Estimated percentages (by volume) of the most common minerals in Earth's crust.**

NOTES:

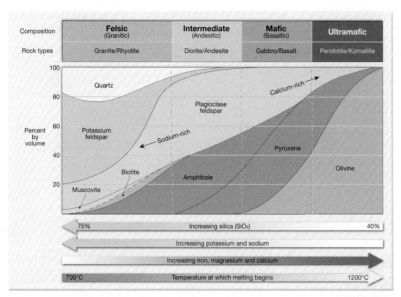

Figure 3.10 Mineralogy of common igneous rocks and the magmas from which they form.

Chemical Composition	Felsic (Granitic)	Intermediate (Andesitic)	Mafic (Basaltic)	Ultramafic
Dominant Minerals	Quartz Potassium feldspar Sodium-rich plagioclase feldspar	Amphibole Sodium- and calcium-rich plagioclase feldspar	Pyroxene Calcium-rich plagioclase feldspar	Olivine Pyroxene
Accessory Minerals	Amphibole Muscovite Biotite	Pyroxene Biotite	Amphibole Olivine	Calcium-rich plagioclase feldspar
Phaneritic (coarse-grained)	Granite	Diorite	Gabbro	Peridotite
Aphanitic (fine-grained)	Rhyolite	Andesite	Basalt	Komatiite (rare)
Porphyritic	"Porphyritic" precedes any of the above names whenever there are appreciable phenocrysts			Uncommon
Glassy	Obsidian (compact glass) Pumice (frothy glass)			
Pyroclastic (fragmental)	Tuff (fragments less than 2 mm) Volcanic Breccia (fragments greater than 2 mm)			
Rock Color (based on % of dark minerals)	0% to 25%	25% to 45%	45% to 85%	85% to 100%

Figure 3.11 Classification of the major groups of igneous rocks.

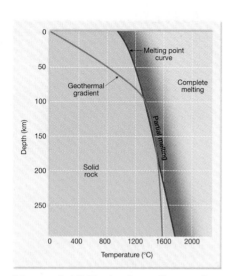

Figure 3.18 **Typical geothermal gradient.**

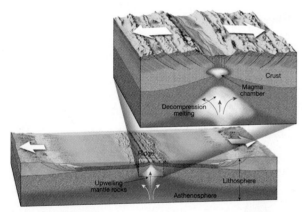

Figure 3.19 **A drop in confining pressure can trigger melting.**

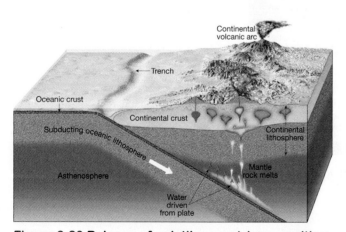

Figure 3.20 **Release of volatiles can trigger melting.**

NOTES:

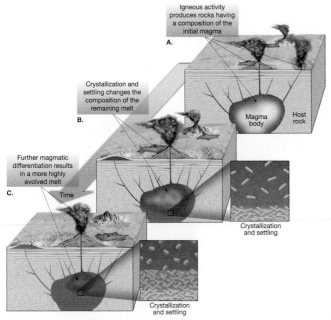

Figure 3.21 Bowen's reaction series.

Figure 3.22 How a magma evolves.

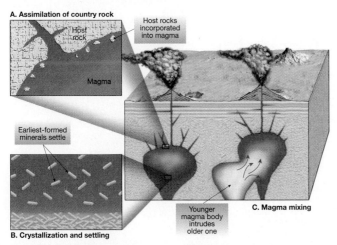

Figure 3.23 Three ways the composition of a magma body may be altered.

NOTES:

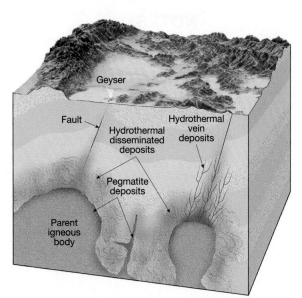

Figure 3.26 Relationship between a parent igneous body and the associated pegmatite and hydrothermal deposits.

NOTES:

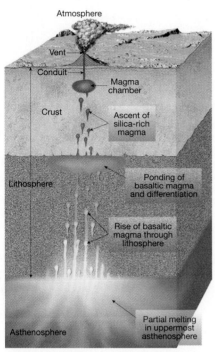

Figure 4.4 Movement of magma from its source through the continental crust.

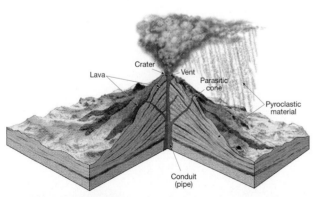

Figure 4.10 Anatomy of a "typical" composite cone.

NOTES:

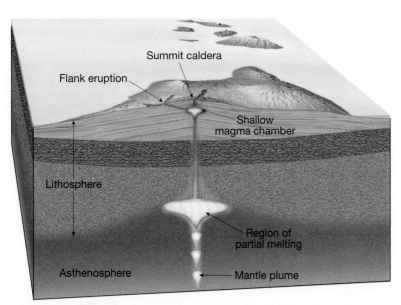

Figure 4.12 Mauna Loa is one of five shield volcanoes that make up Hawaii.

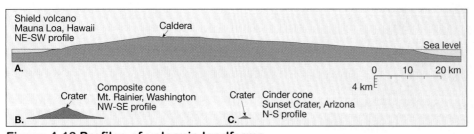

Figure 4.13 Profiles of volcanic landforms.

NOTES:

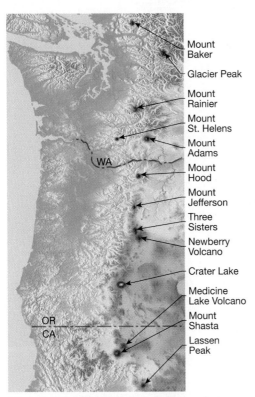

Figure 4.18 Potentially active volcanoes in the Cascade Range.

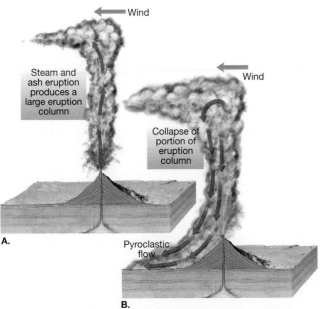

Figure 4.20A,B Develoment of a pyroclastic flow.

NOTES:

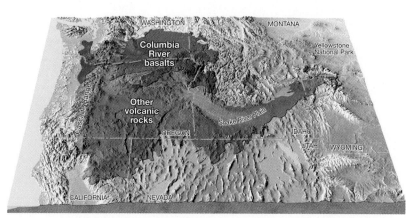

Figure 4.23 **Volcanic areas that comprise the Columbia Plateau.**

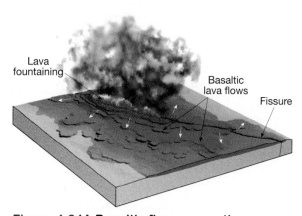

Figure 4.24A **Basaltic fissure eruption.**

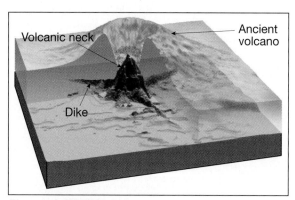

Figure 4.27 **Volcanic neck.**

NOTES:

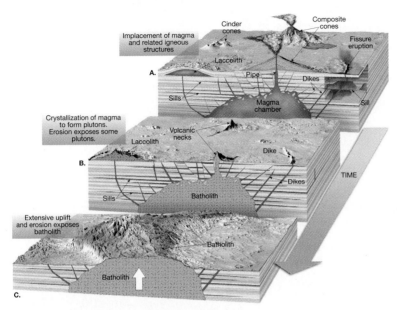

Figure 4.28 **Basic igneous structures.**

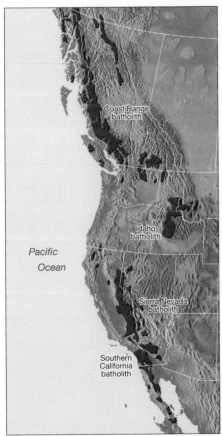

Figure 4.32 **Granitic batholits along the western margin of North America.**

NOTES:

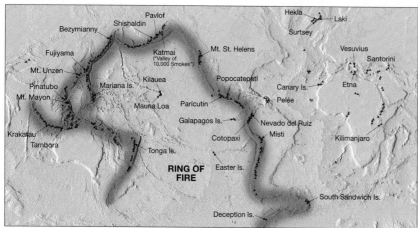

Figure 4.33 Locations of some of Earth's major volcanoes.

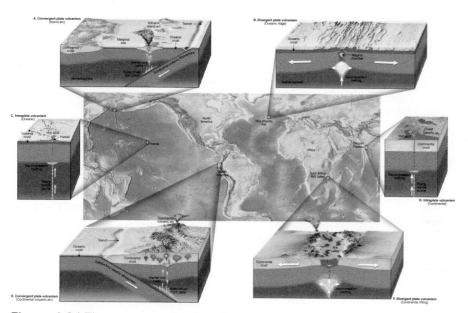

Figure 4.34 Three zones of volcanism.

NOTES:

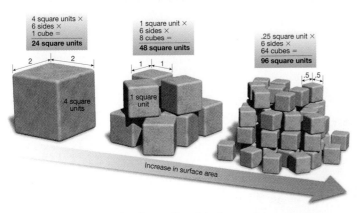

Figure 5.3 Mechanical weathering increases surface area for chemical weathering.

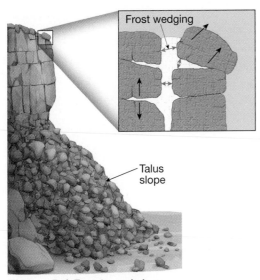

Figure 5.4 Frost wedging.

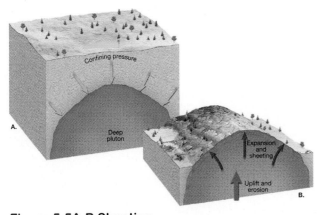

Figure 5.5A,B Sheeting.

NOTES:

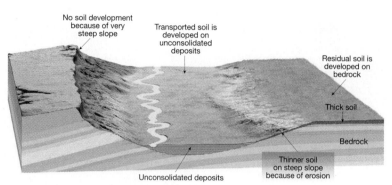

Figure 5.15 Residual and transported soils.

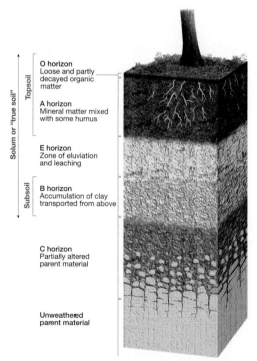

Figure 5.18 Idealized soil profile from a humid climate in the middle latitudes.

NOTES:

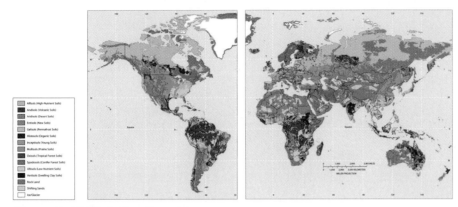

Figure 5.19 Global soil regions.

NOTES:

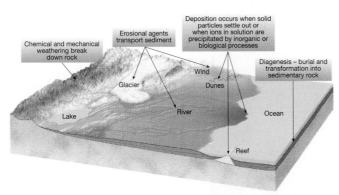

Figure 6.2 Portion of the rock cycle that pertains to the formation of sedimentary rocks.

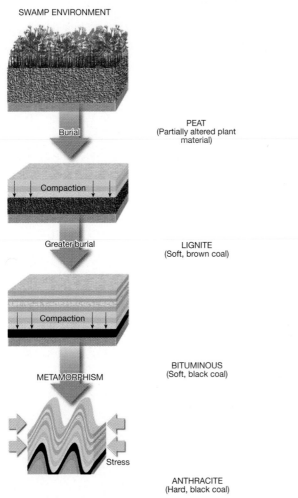

Figure 6.17 Formation of coal.

NOTES:

Detrital Sedimentary Rocks				Chemical and Organic Sedimentary Rocks			
ClasticTexture (particle size)		Sediment Name	Rock Name	Composition	Textur	Rock Name	
Coarse (over 2 mm)		Gravel (Rounded particles)	Conglomerate	Calcite, CaCO$_3$	Nonclastic: Fine to coarse crystalline	Crystalline Limestone	
		Gravel (Angular particles)	Breccia			Travertine	
Medium (1/16 to 2 mm)		Sand (If abundant feldspar is present the rock is called **Arkose**)	Sandstone		Clastic: Visible shells and shell fragments loosely cemented	Coquina	B i o c h e m i c a l L i m e s t o n e
					Clastic: Various size shells and shell fragments cemented with calcite cement	Fossiliferous Limestone	
Fine (1/16 to 1/256 mm)		Mud	Siltstone		Clastic: Microscopic shells and clay	Chalk	
Very fine (less than 1/256 mm)		Mud	Shale or Mudstone	Quartz, SiO$_2$	Nonclastic: Very fine crystalline	Chert (light colored) Flint (dark colored)	
				Gypsum CaSO$_4$•2H$_2$O	Nonclastic: Fine to coarse crystalline	Rock Gypsum	
				Halite, NaCl	Nonclastic: Fine to coarse crystalline	Rock Salt	
				Altered plant fragments	Nonclastic: Fine-grained organic matter	Bituminous Coal	

Figure 6.18 Identification of sedimentary rocks.

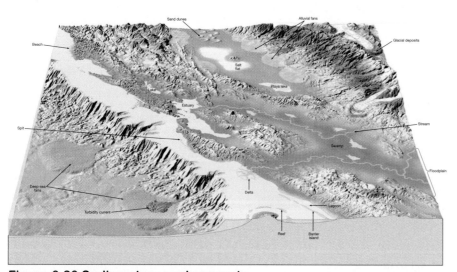

Figure 6.20 Sedimentary environments.

NOTES:

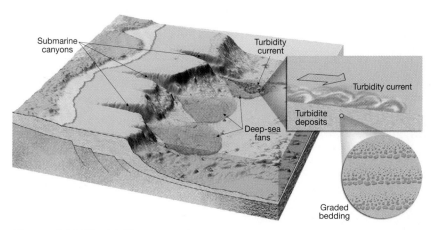

Figure 6.23 Turbidity currents.

Figure 6.25 Annual per capita consumption of nonmetallic and metallic mineral resources for the United States.

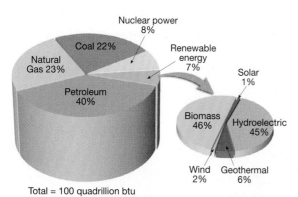

Total = 100 quadrillion btu

Figure 6.26 U.S. energy consumption, 2004.

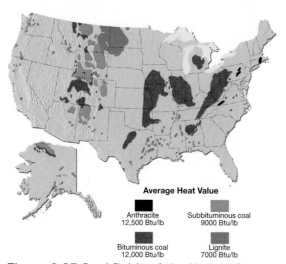

Figure 6.27 Coal fields of the United States.

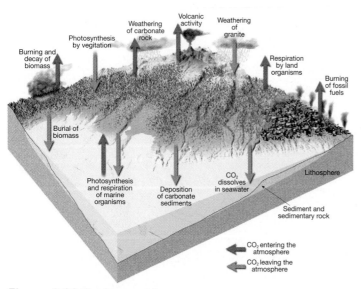

Figure 6.28 Carbon cycle.

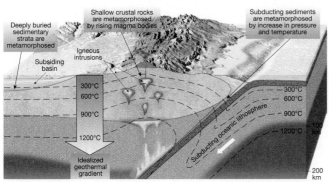

Figure 7.2 Geothermal gradient and its role in metamorphism.

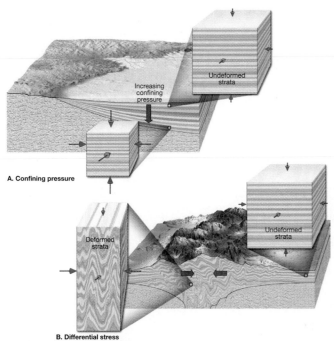

Figure 7.3 Confining pressure and differential stress
as metamorphic agents.

NOTES:

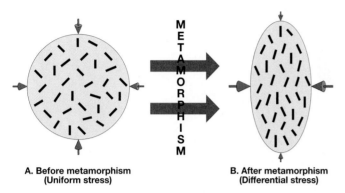

A. Before metamorphism
(Uniform stress)

B. After metamorphism
(Differential stress)

Figure 7.5 Mechanical rotation of platy or elongated mineral grains.

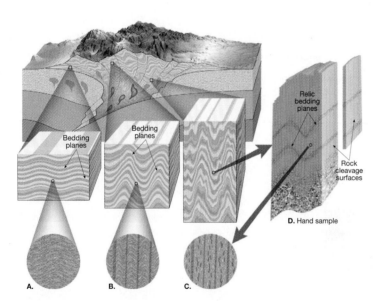

Figure 7.6 Development of one type of rock cleavage.

NOTES:

Rock Name	Texture		Grain Size	Comments	Original Parent Rock
Slate	Foliated		Very fine	Excellent rock cleavage, smooth dull surfaces	Shale, mudstone, or siltstone
Phyllite			Fine	Breaks along wavy surfaces, glossy sheen	Shale, mudstone, or siltstone
Schist			Medium to Coarse	Micaceous minerals dominate, scaly foliation	Shale, mudstone, or siltstone
Gneiss			Medium to Coarse	Compositional banding due to segregation of minerals	Shale, granite, or volcanic rocks
Migmatite			Medium to Coarse	Banded rock with zones of light-colored crystalline minerals	Shale, granite, or volcanic rocks
Mylonite	Weakly Foliated		Fine	When very fine-grained, resembles chert, often breaks into slabs	Any rock type
Metaconglomerate			Coarse-grained	Stretched pebbles with preferred orientation	Quartz-rich conglomerate
Marble	Nonfoliated		Medium to coarse	Interlocking calcite or dolomite grains	Limestone, dolostone
Quartzite			Medium to coarse	Fused quartz grains, massive, very hard	Quartz sandstone
Hornfels			Fine	Usually, dark massive rock with dull luster	Any rock type
Anthracite			Fine	Shiny black rock that may exhibit conchoidal fracture	Bituminous coal
Fault breccia			Medium to very coarse	Broken fragments in a haphazard arrangement	Any rock type

(Increasing Metamorphism — arrow pointing downward alongside the Foliated rocks)

Figure 7.10 Classification of common metamorphic rocks.

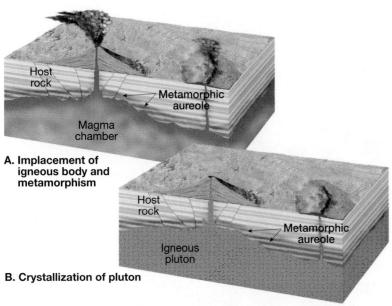

A. Implacement of igneous body and metamorphism

B. Crystallization of pluton

Figure 7.13A,B Contact metamorphism produces a zone of alteration called an aureole around an intrusive igneous body.

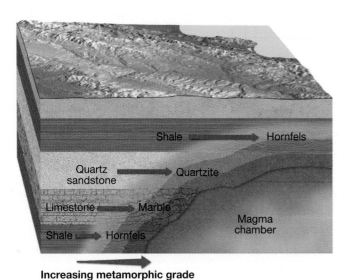

Increasing metamorphic grade

Figure 7.14 Contact metamorphism of different rock types.

NOTES:

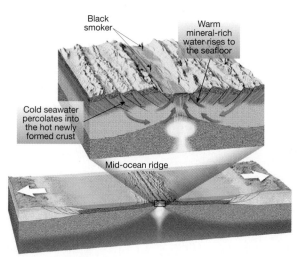

Figure 7.16 Hydrothermal metamorphism along a mid-ocean ridge.

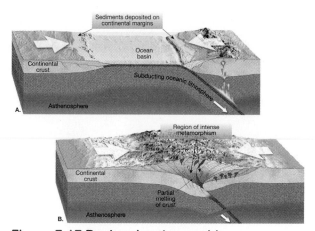

Figure 7.17 Regional metamorphism.

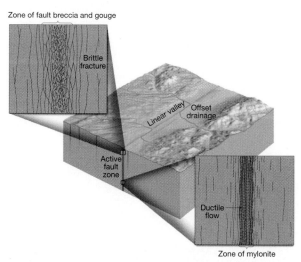

Figure 7.18 Metamorphism along a fault zone.

NOTES:

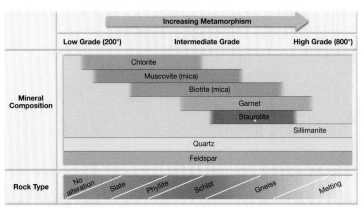

Figure 7.20 Typical transition in mineralogy resulting from progressive metamorphism of shale.

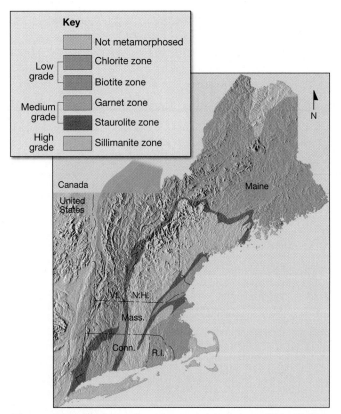

Figure 7.21 Zones of metamorphic intensities in New England.

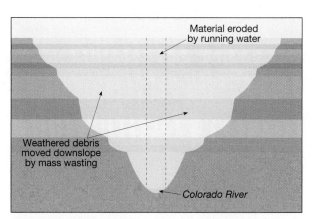

Figure 8.2 Mass-wasting in the Grand Canyon.

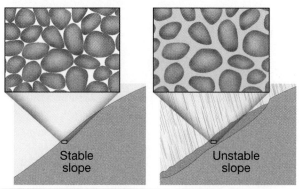

A. Dry soil–high friction   B. Saturated soil

Figure 8.4 Effect of water on mass wasting.

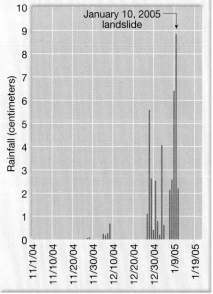

Figure 8.8 Daily rainfall leading up to the January 2005 La Conchita landslide.

Lutgens/Tarbuck, *Essentials of Geology*, 10e
© 2009 Pearson Prentice Hall, Inc.

NOTES:

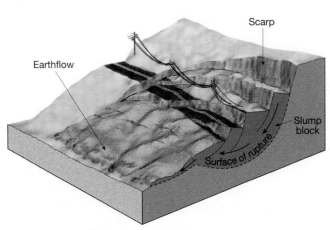

Figure 8.11A **Slump.**

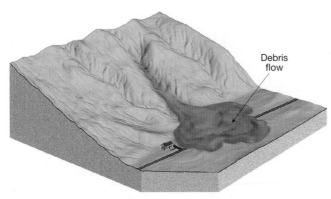

Figure 8.14 **Debris flow.**

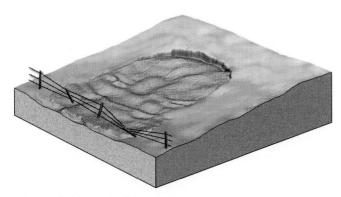

Figure 8.15 **Earthflow.**

NOTES:

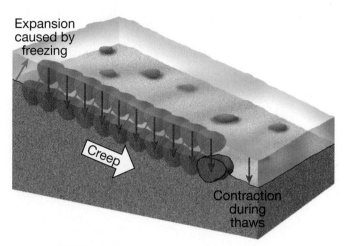

Figure 8.16 Creep.

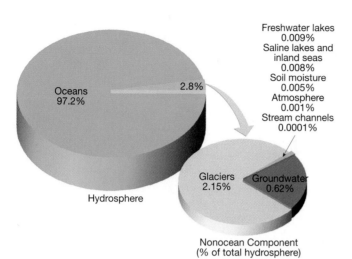

Figure 9.2 **Distribution of Earth's water.**

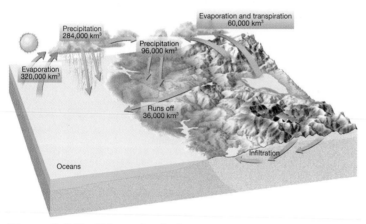

Figure 9.3 **Earth's water balance.**

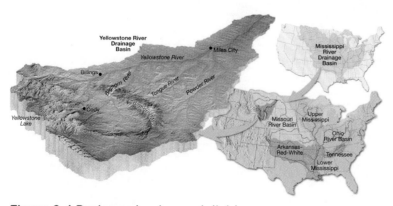

Figure 9.4 **Drainage basins and divides.**

NOTES:

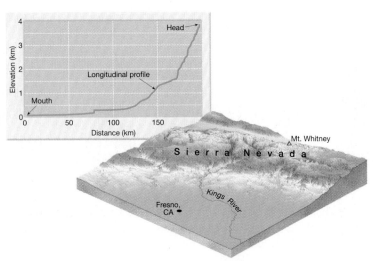

Figure 9.7 **Longitudinal profile of California's King River.**

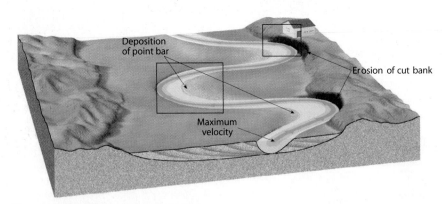

Figure 9.10 **Meandering stream.**

NOTES:

Figure 9.11 Formation of a cutoff and oxbow lake.

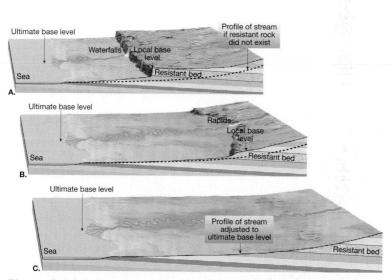

Figure 9.14 A resistant layer of rock can act as a local base level.

NOTES:

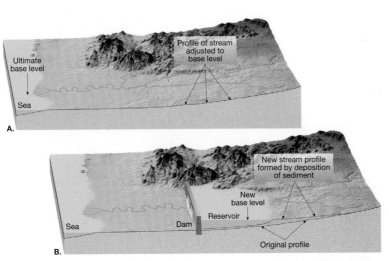

Profile of stream adjusted to base level

Ultimate base level

Sea

A.

New stream profile formed by deposition of sediment

New base level

Reservoir

Sea   Dam

Original profile

B.

Figure 9.15 **Dam and reservoir.**

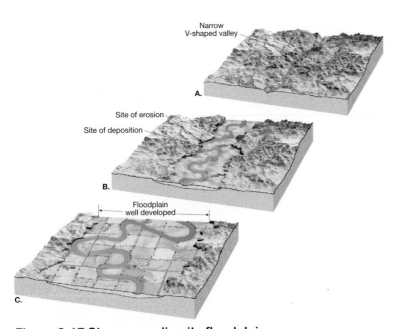

Narrow V-shaped valley

A.

Site of erosion

Site of deposition

B.

Floodplain well developed

C.

Figure 9.17 **Stream eroding its floodplain.**

NOTES:

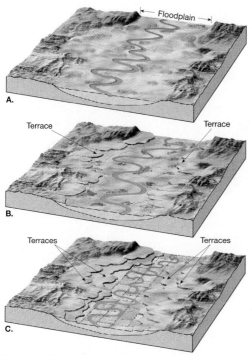

Figure 9.18A,B,C Development of multiple stream terraces.

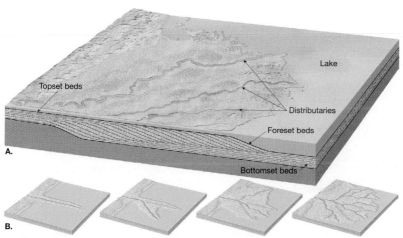

Figure 9.20 Structure and growth of a simple delta.

NOTES:

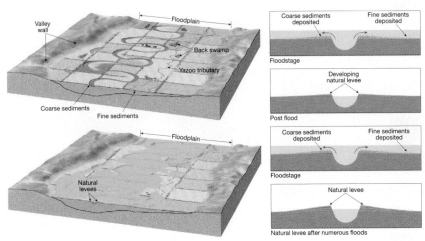

Figure 9.22 **Natural levees.**

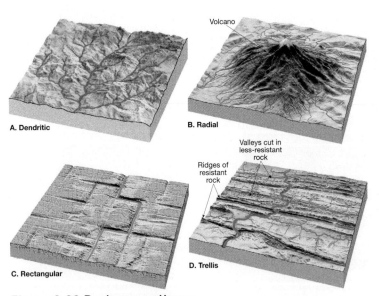

Figure 9.23 **Drainage patterns.**

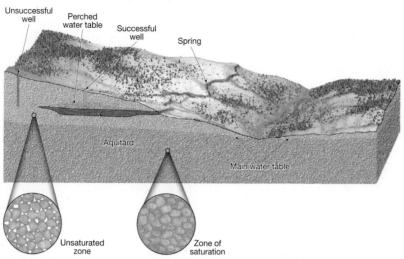

NOTES:

Figure 10.3 Features associated with subsurface water.

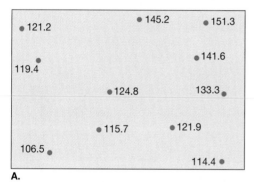

A.

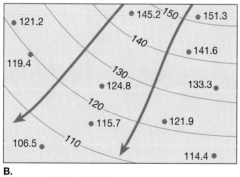

B.

EXPLANATION

• Location of well and elevation of water
   table above sea level, in feet

~120~ Water table contour shows elevation of
   water table, contour interval 10 feet

◄— Ground-water flow line

Figure 10.4 Water table map.

NOTES:

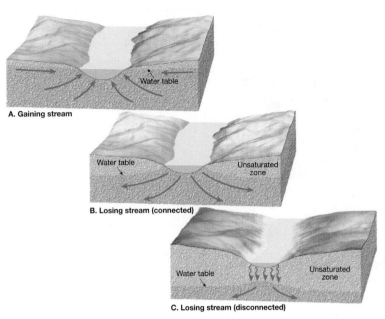

Figure 10.5 Interaction between the groundwater system and streams.

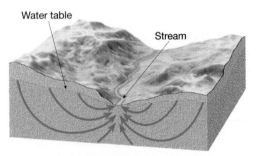

Figure 10.6 Groundwater movement through uniformly permeable material.

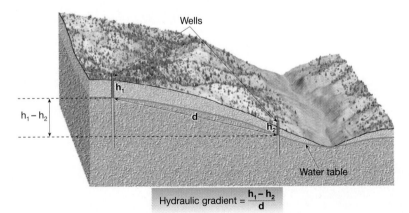

Figure 10.7 Hydraulic gradient.

Lutgens/Tarbuck, *Essentials of Geology*, 10e
© 2009 Pearson Prentice Hall, Inc.

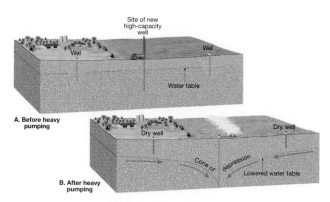

Figure 10.10 Cone of depression.

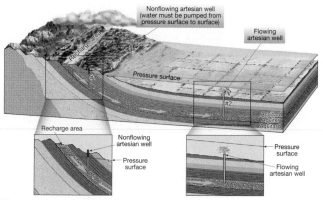

Figure 10.11 Artesian systems.

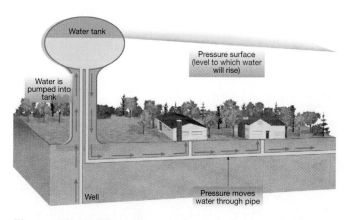

Figure 10.13 City water systems can be considered artificial artesian systems.

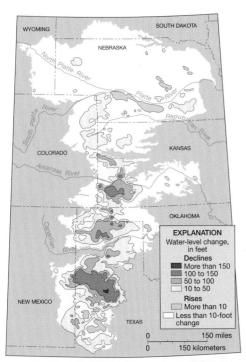

Figure 10.14 Changes in groundwater levels in the High Plains aquifer through time.

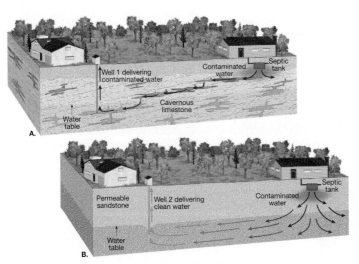

Figure 10.16 Contamination of wells.

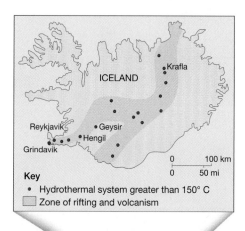

Figure 10.21 Hydrothermal activity in Iceland.

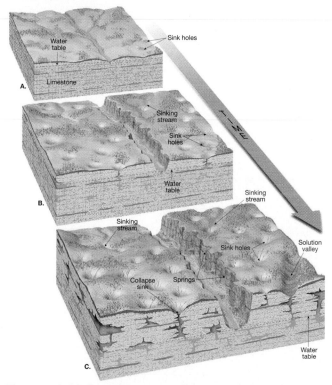

Figure 10.25 Development of a karst landscape.

NOTES:

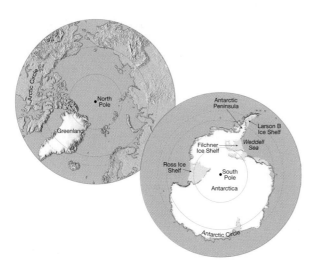

Figure 11.2 Present day continental ice sheets.

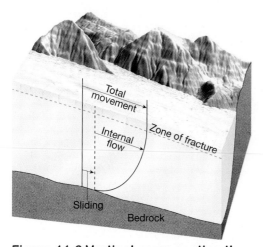

Figure 11.6 Vertical cross section through a glacier.

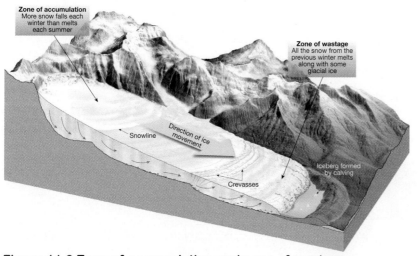

Figure 11.9 Zone of accumulation and zone of wastage.

NOTES:

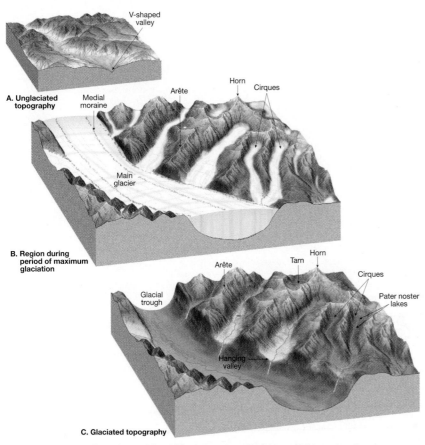

Figure 11.13 **Erosional landforms created by alpine glaciers.**

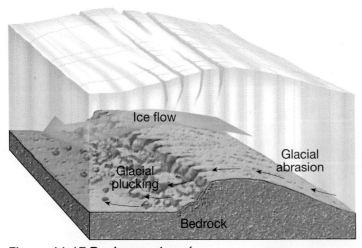

Figure 11.17 **Roche moutonnée.**

NOTES:

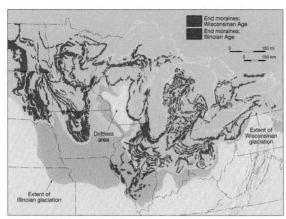

Figure 11.21 **End moraines of the Great Lakes region.**

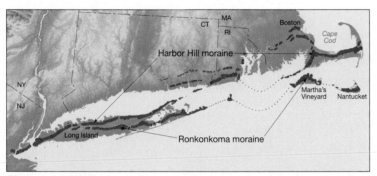

Figure 11.22 **End morinaes make up substantial parts of Long Island, Cape Cod, Martha's Vineyard, and Nantucket.**

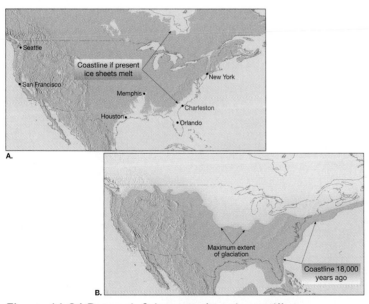

Figure 11.24 **Present, future, and past coastline.**

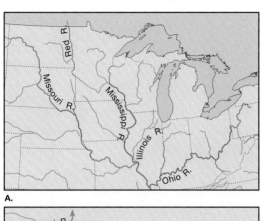

A.

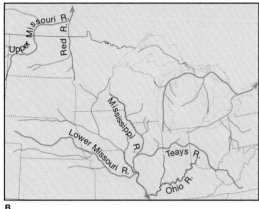

B.

Figure 11.25 Great lakes region, present and past.

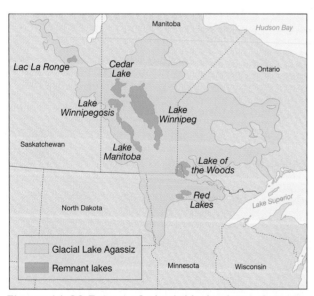

Figure 11.26 Extent of glacial Lake Agassiz.

NOTES:

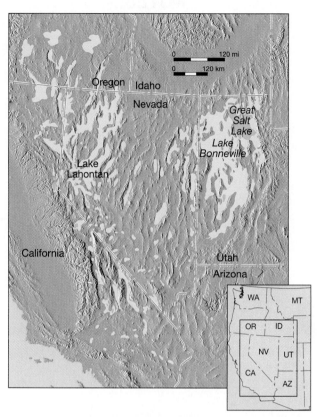

Figure 11.27 Pluvial lakes of the western United States.

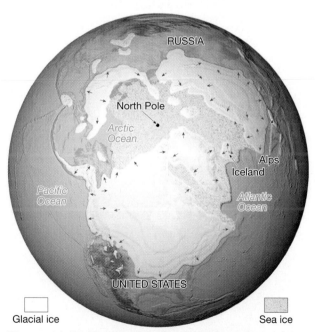

Figure 11.28 Maximum extent of glaciation in the Northern Hemisphere during the Ice Age.

Lutgens/Tarbuck, *Essentials of Geology*, 10e
© 2009 Pearson Prentice Hall, Inc..

NOTES:

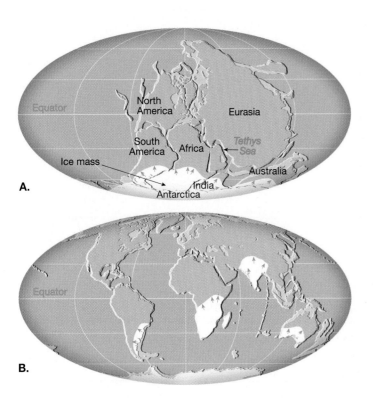

A.

B.

Figure 11.29 Location of ancient glacial ice sheets.

NOTES:

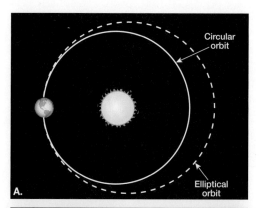

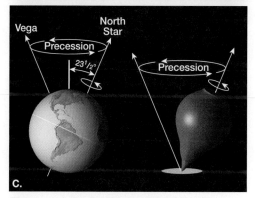

Figure 11.30 Orbital variations.

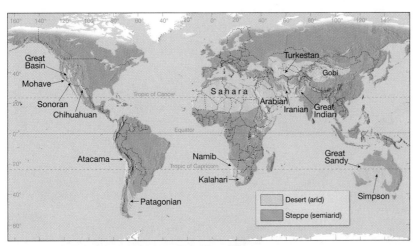

Figure 12.2 Arid and semiarid climates.

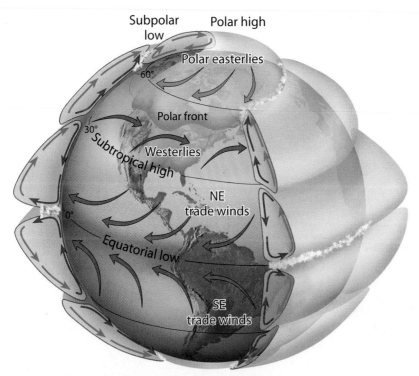

Figure 12.3A Idealized diagram of Earth's general circulation.

NOTES:

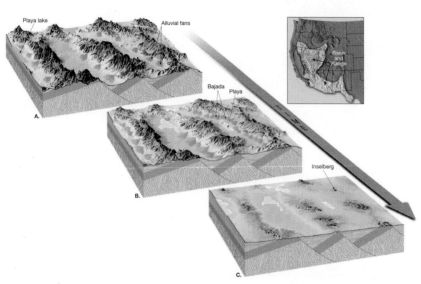

Figure 12.7 Stages of landscape evolution in a mountainous desert.

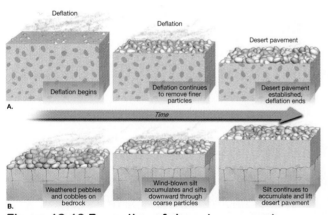

Figure 12.13 Formation of desert pavement.

NOTES:

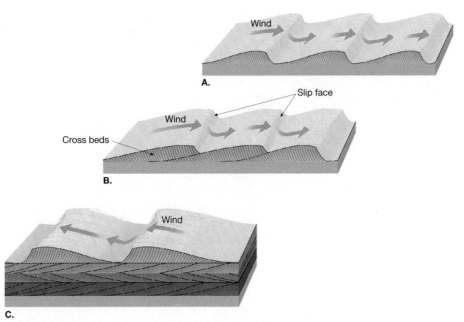

Figure 12.17A,B,C Dunes and formation of cross beds.

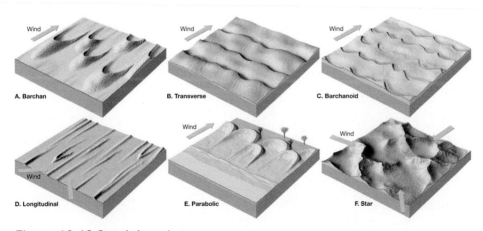

Figure 12.18 Sand dune types.

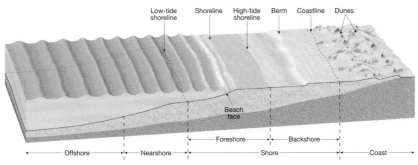

Figure 13.3 Coastal zone.

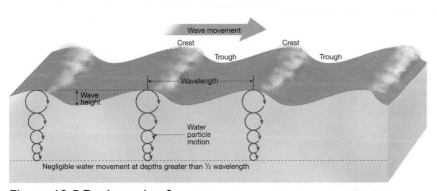

Figure 13.5 Basic parts of a wave.

NOTES:

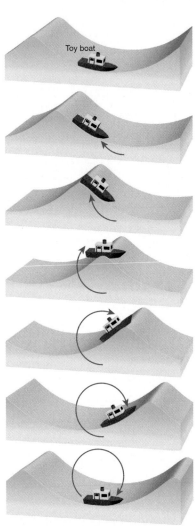

Figure 13.6 Wave form advances while water does not.

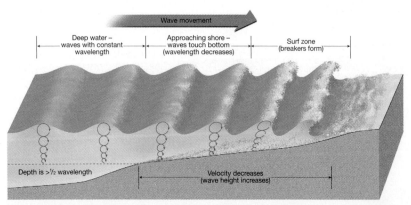

Figure 13.7 Changes that occur when a wave moves onto shore.

NOTES:

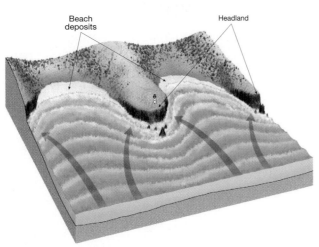

Figure 13.11 Wave refraction along an irregular coastline.

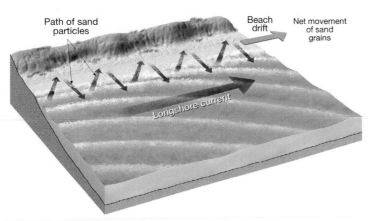

Figure 13.12 Beach drift and longshore current.

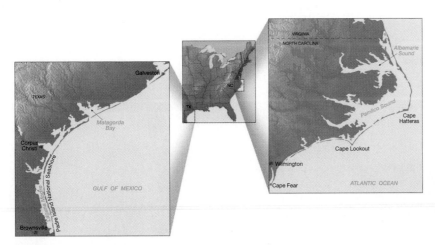

Figure 13.16 Barrier islands.

NOTES:

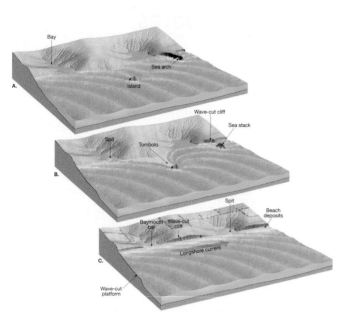

Figure 13.17 Changes through time along an initially irregular coastline.

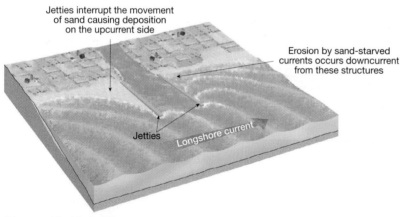

Figure 13.18 Jetties.

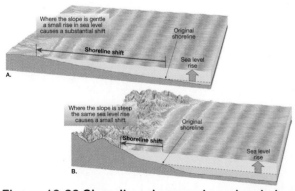

Figure 13.23 Shoreline slope and sea-level change.

NOTES:

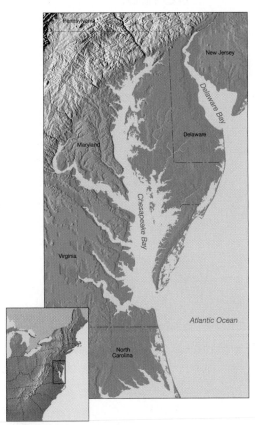

Figure 13.27 Estuaries along the East Coast of the United States.

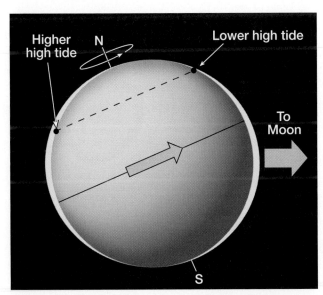

Figure 13.29 Idealized tidal bulges on Earth.

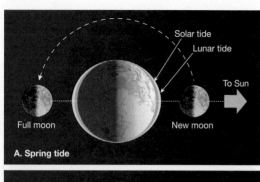

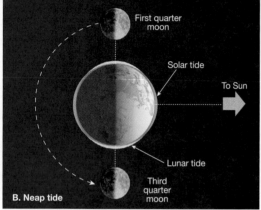

Figure 13.30 Earth-Moon-Sun positions and the tides.

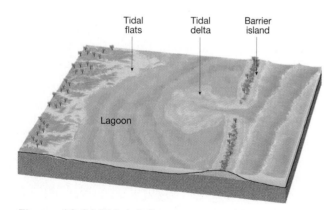

Figure 13.31 Tidal delta.

NOTES:

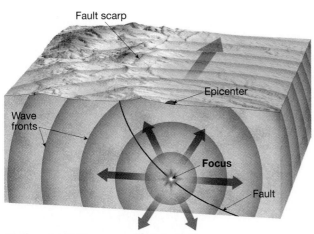

Figure 14.2 Focus and epicenter.

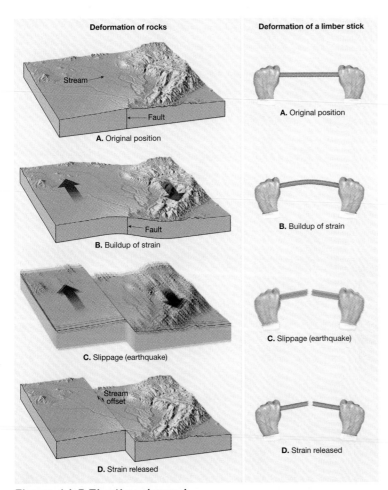

Figure 14.5 Elastic rebound.

NOTES:

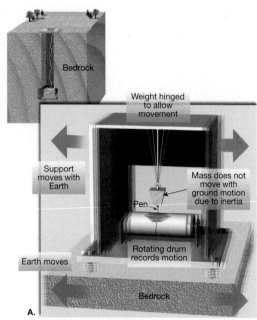

Figure 14.7A Principle of the seismograph.

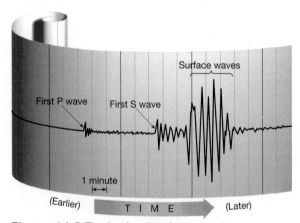

Figure 14.8 Typical seismic record.

NOTES:

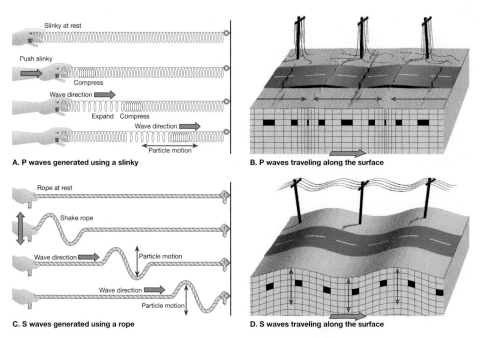

A. P waves generated using a slinky

B. P waves traveling along the surface

C. S waves generated using a rope

D. S waves traveling along the surface

Figure 14.9 Types of seismic waves and their characteristic motion.

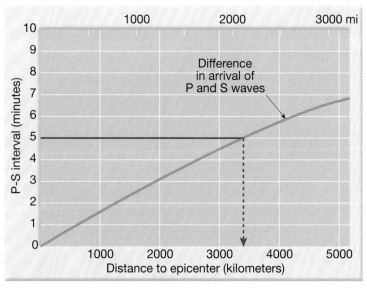

Figure 14.10 Travel-time graph.

NOTES:

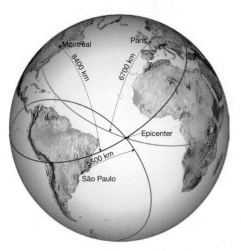

Figure 14.11 **Locating the epicenter.**

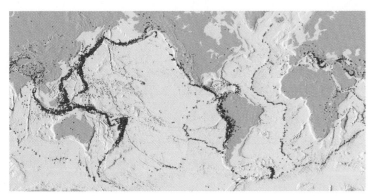

Figure 14.12 **Distribution of earthquakes with magnitudes greater than 5 for a 10-year period.**

Figure 14.13 **Zones of destruction from 1925 earthquake in Japan.**

Lutgens/Tarbuck, *Essentials of Geology*, 10e
© 2009 Pearson Prentice Hall, Inc.

NOTES:

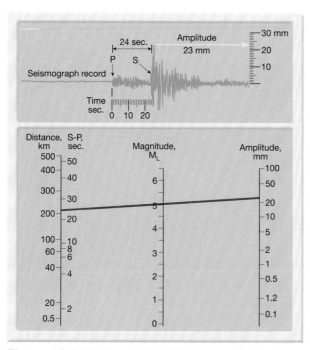

Figure 14.14 Determining Richter magnitude.

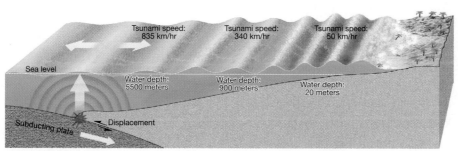

Figure 14.18 Tsunami generated by displacement of the ocean floor.

NOTES:

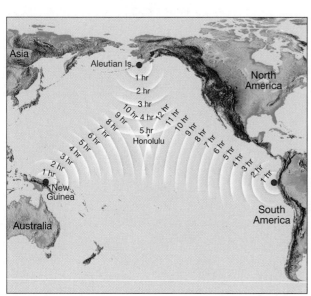

Figure 14.20 Tsunami travel times to Honolulu, Hawaii.

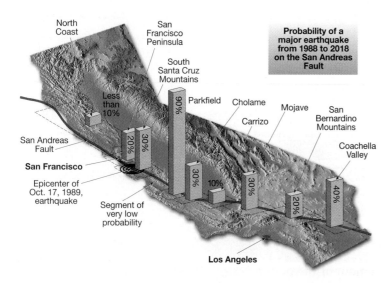

Figure 14.24 Probability of a major earthquake from 1988 to 2018 on the San Andreas fault.

NOTES:

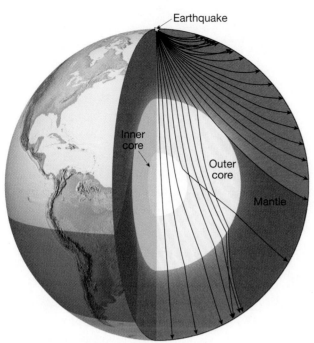

Figure 14.25 Paths of seismic rays.

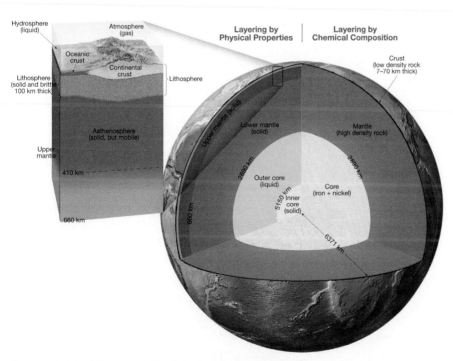

Figure 14.26 Views of Earth's layered structure.

NOTES:

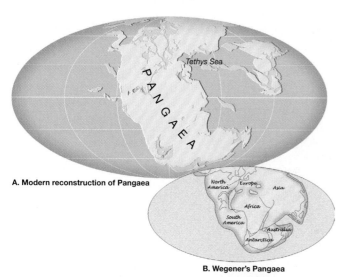

Figure 15.2 **Reconstructions of Pangaea.**

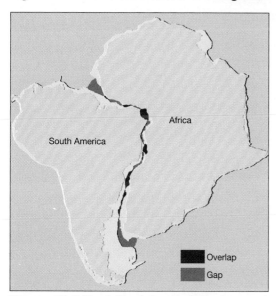

Figure 15.3 **Best fit of South America and Africa.**

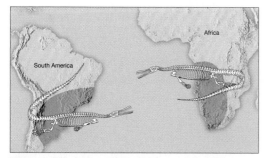

Figure 15.4 **Fossils of** *Mesosaurus*.

NOTES:

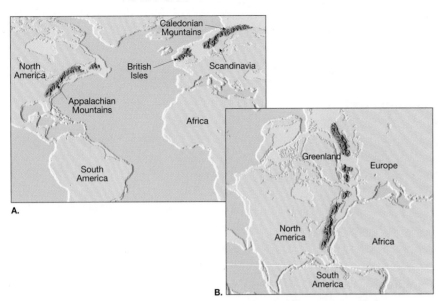

Figure 15.6 Matching mountain ranges across the North Atlantic.

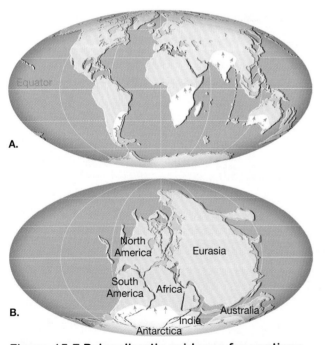

Figure 15.7 Paleoclimatic evidence for continental drift.

NOTES:

Figure 15.8 Illustration of some of Earth's lithospheric plates.

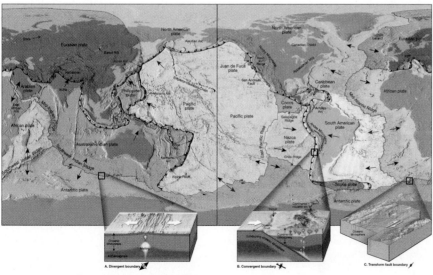

Figure 15.9 Earth's tectonic plates.

NOTES:

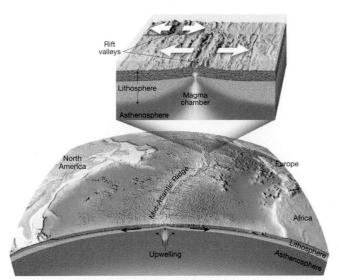

Figure 15.10 Most divergent plate boundaries are along the crests of oceanic ridges.

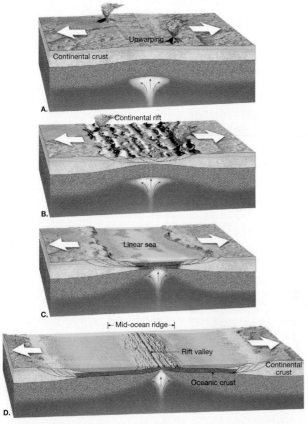

Figure 15.11 Continental rifting and the formation of a new ocean basin.

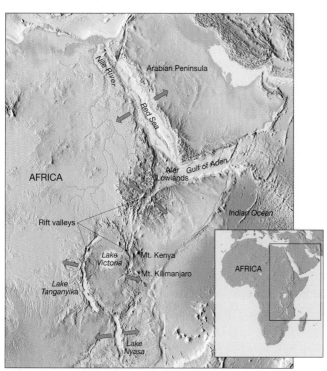

Figure 15.12 **East African rift valleys.**

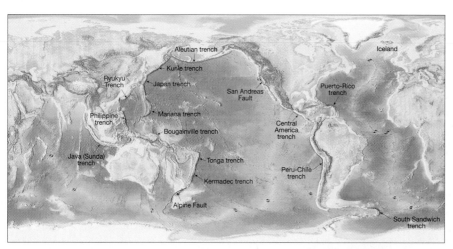

Figure 15.13 **Global distribution of trenches, the ridge system, and transform faults.**

NOTES:

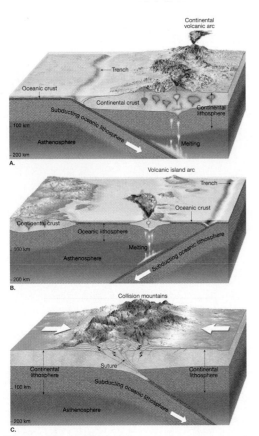

Figure 15.14 Three types of convergent plate boundaries.

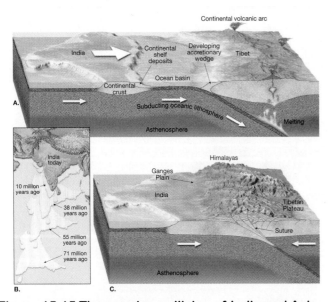

Figure 15.15 The ongoing collision of India and Asia produced the Himalayas.

NOTES:

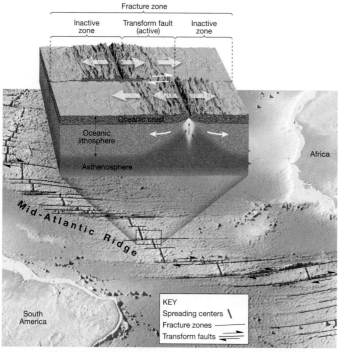

Figure 15.16 **Transform fault joining segments of the Mid-Atlantic Ridge.**

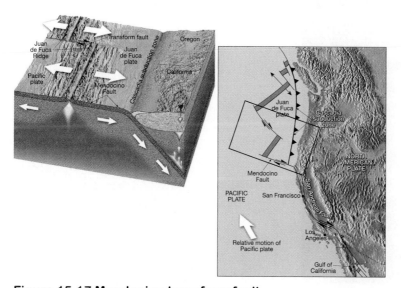

Figure 15.17 **Mendocino transform fault.**

NOTES:

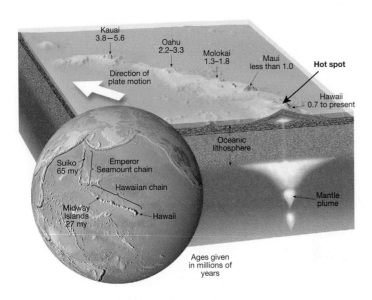

Figure 15.18 Chain of islands and seamounts resulting from the movement of the Pacific plate over an apparently stationary hot spot.

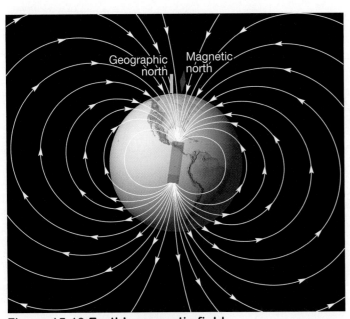

Figure 15.19 Earth's magnetic field.

NOTES:

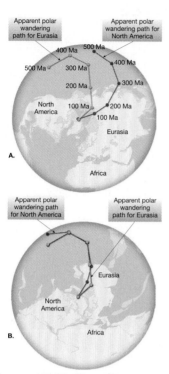

Figure 15.20 Simplified apparent polar wandering paths.

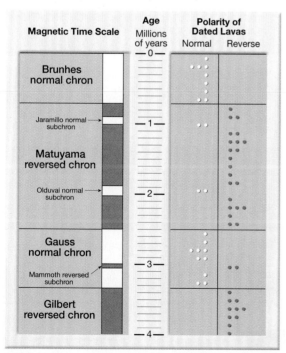

Figure 15.21 Time scale of Earth's magnetic field.

NOTES:

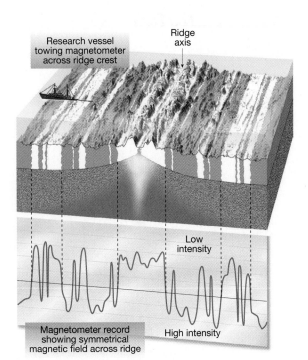

Figure 15.22 **The ocean floor as a magnetic tape recorder.**

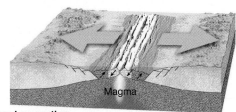

A. Period of normal magnetism

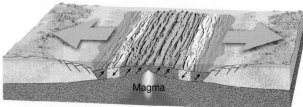

B. Period of reverse magnetism

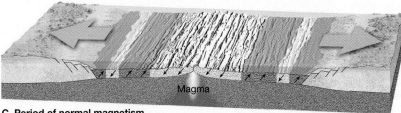

C. Period of normal magnetism

Figure 15.23 **As new basalt is added to the ocean floor, it is magnetized according to Earth's existing magnetic field.**

NOTES:

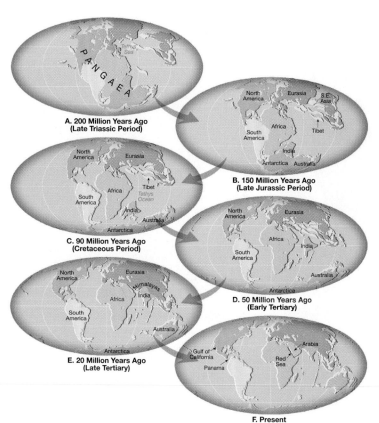

Figure 15.24 **Several views of the breakup of Pangaea.**

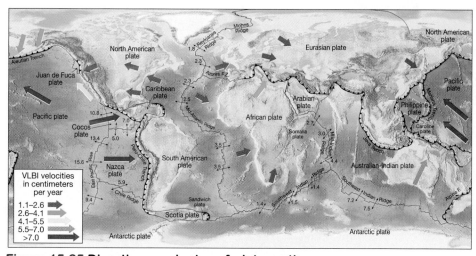

Figure 15.25 **Directions and rates of plate motion.**

NOTES:

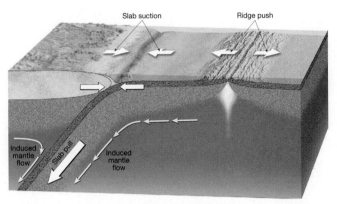

Figure 15.26 **Some of the forces that act on plates.**

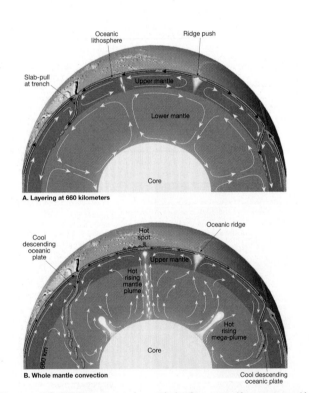

Figure 15.27 **Proposed models for mantle convection.**

NOTES:

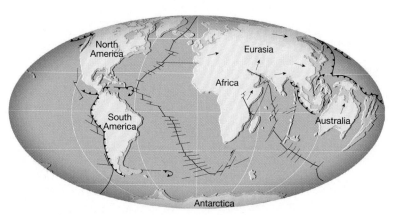

Figure 15.28 The world as it may look 50 million years from now.

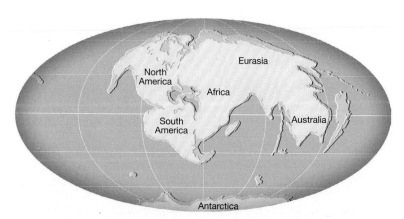

Figure 15.29 The world as it may look 250 million years from now.

NOTES:

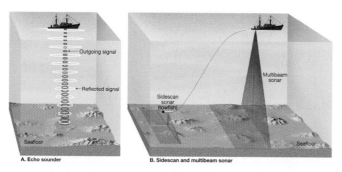

Figure 16.2 Various types of sonar.

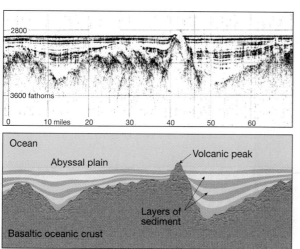

Figure 16.3 Seismic cross section and matching sketch across a portion of the Madeira abyssal plain.

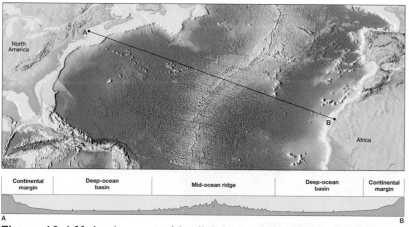

Figure 16.4 Major topographic divisions of the North Atlantic.

NOTES:

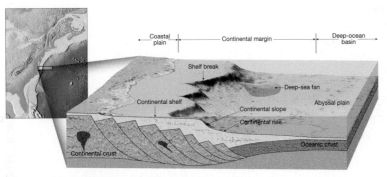

Figure 16.5 Passive continental margin.

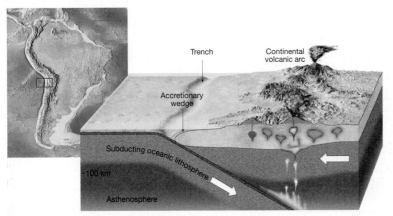

Figure 16.6 Active continental margin.

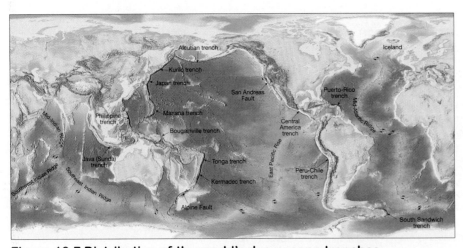

Figure 16.7 Distribution of the world's deep-ocean trenches.

NOTES:

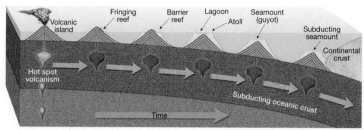

Figure 16.9 **Formation of a coral atoll.**

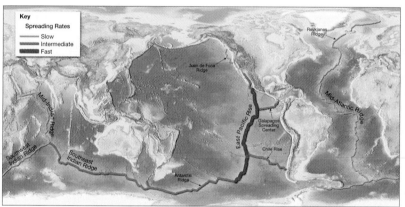

Figure 16.10 **Distribution of the oceanic ridge system.**

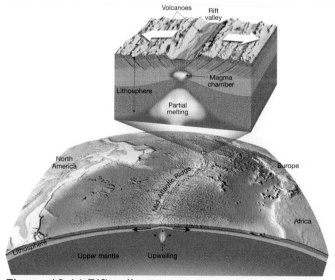

Figure 16.11 **Rift valleys.**

NOTES:

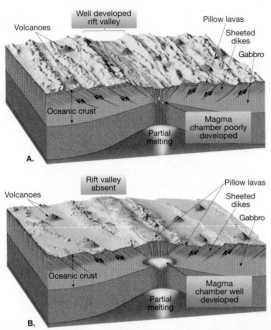

Figure 16.12 Topography of the crest of an oceanic ridge.

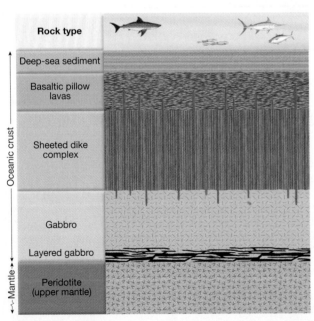

Figure 16.14 Rock types associated with a typical section of oceanic crust.

NOTES:

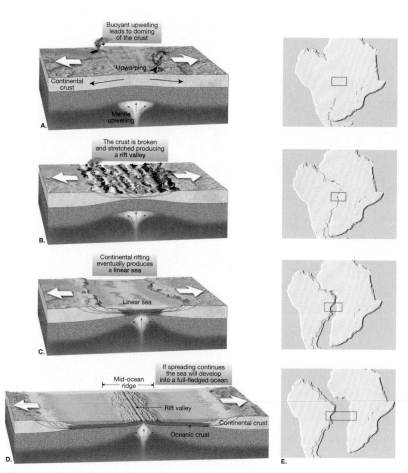

Figure 16.17 Formation of an ocean basin.

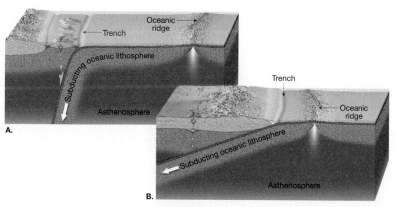

Figure 16.18 Subduction angle.

NOTES:

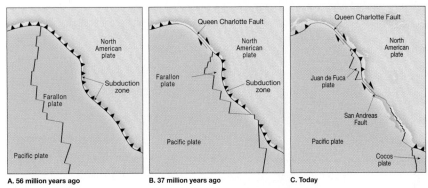

A. 56 million years ago   B. 37 million years ago   C. Today

Figure 16.19 Demise of the Farallon plate.

NOTES:

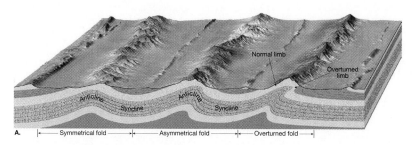

Figure 17.3A Principal types of folded strata.

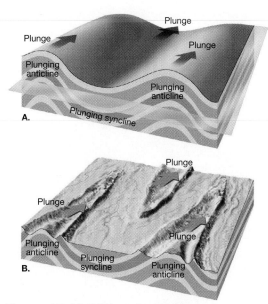

Figure 17.4A,B Plunging folds.

NOTES:

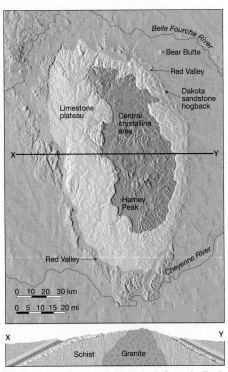

Figure 17.6 Black Hills of South Dakota.

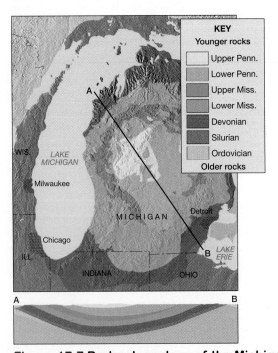

Figure 17.7 Bedrock geology of the Michigan Basin.

NOTES:

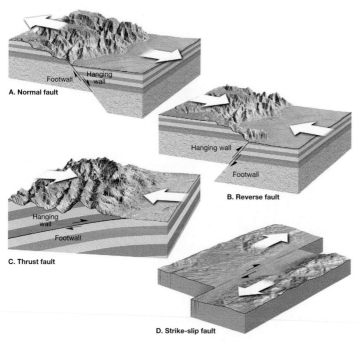

Figure 17.9 **Four types of faults.**

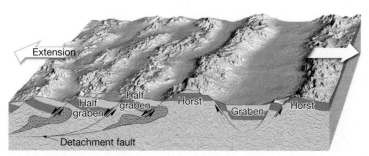

Figure 17.10 **Normal faulting in the Basin and Range Province.**

NOTES:

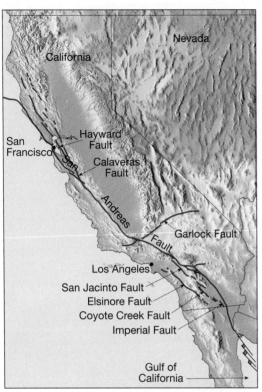

Figure 17.11 San Andreas fault system.

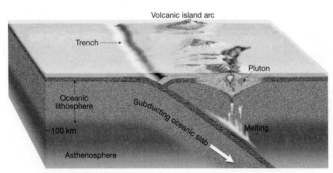

Figure 17.13 Development of a volcanic island arc.

NOTES:

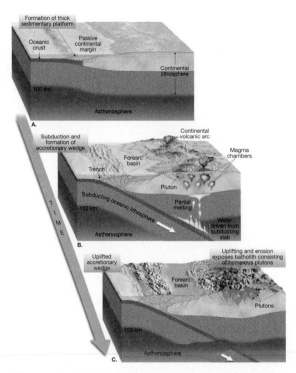

Figure 17.13 Mountain building along an Andean-type subduction zone.

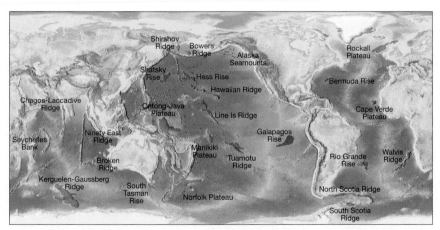

Figure 17.15 Distribution of present-day oceanic plateaus and other submerged crustal fragments.

NOTES:

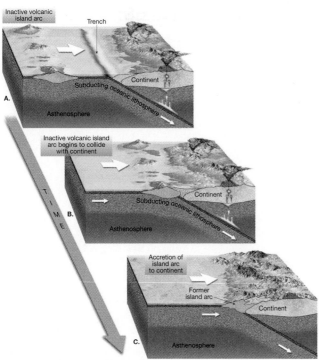

Figure 17.16 Collision of an inactive volcanic island-arc with an Andean-type plate margin.

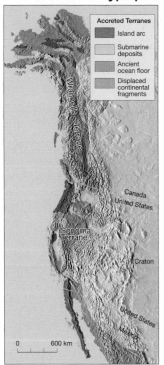

Figure 17.17 Terranes thought to have been added to western North America during the past 200 million years.

NOTES:

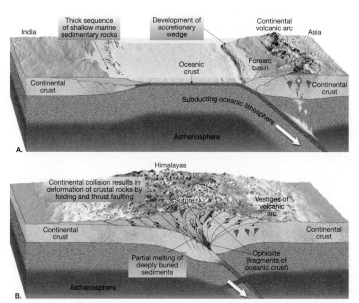

Figure 17.18 Collision of India with the Eurasian plate.

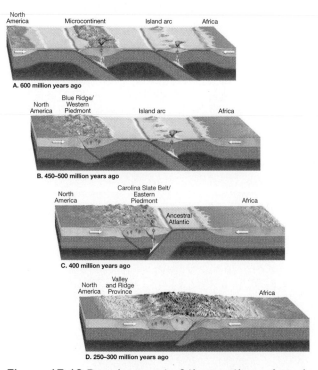

Figure 17.19 Development of the southern Appalachians as the ancient North Atlantic was closed.

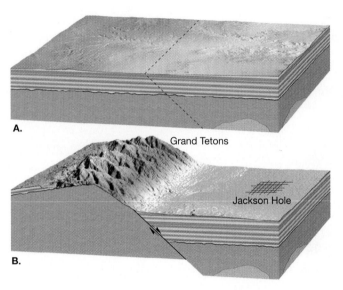

Figure 17.20A,B Fault block mountains.

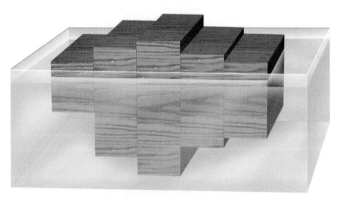

Figure 17.21 How wooden blocks of different thicknesses float in water.

NOTES:

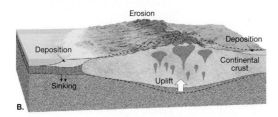

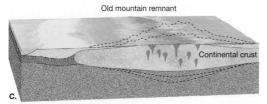

Figure 17.22 Combined effect of erosion and isostatic adjustment.

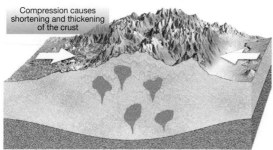

A. Horizontal compressional forces dominate

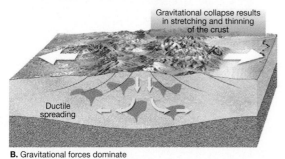

B. Gravitational forces dominate

Figure 17.23 Mountain belt collapsing under its own "weight."

NOTES:

B.
Figure 18.3B Law of superposition in the Grand Canyon.

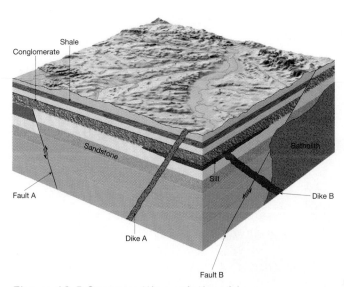

Figure 18.5 Cross-cutting relationships.

NOTES:

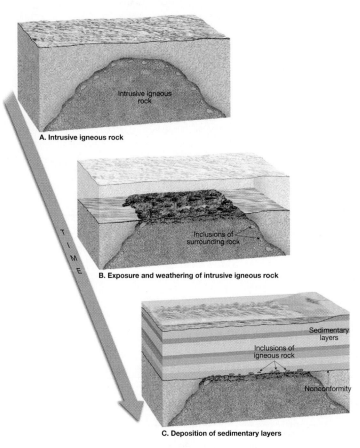

A. Intrusive igneous rock

B. Exposure and weathering of intrusive igneous rock

C. Deposition of sedimentary layers

Figure 18.6 Two ways inclusions can form and non-conformity.

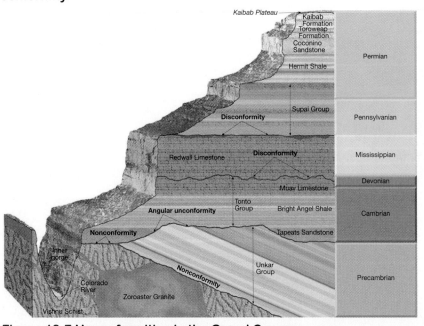

Figure 18.7 Unconformities in the Grand Canyon.

Lutgens/Tarbuck, *Essentials of Geology*, 10e
© 2009 Pearson Prentice Hall, Inc.

NOTES:

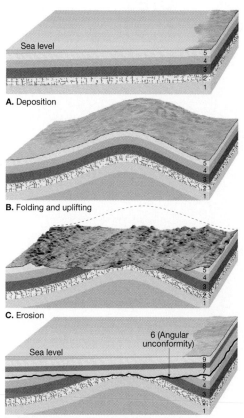

A. Deposition

B. Folding and uplifting

C. Erosion

D. Subsidence and renewed deposition

Figure 18.8 Formation of an angular unconformity.

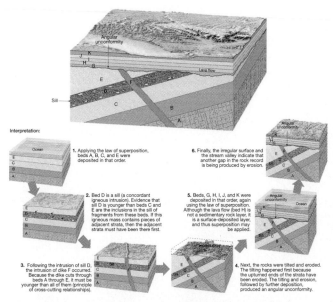

Figure 18.9 Geologic cross section of a hypothetical region.

NOTES:

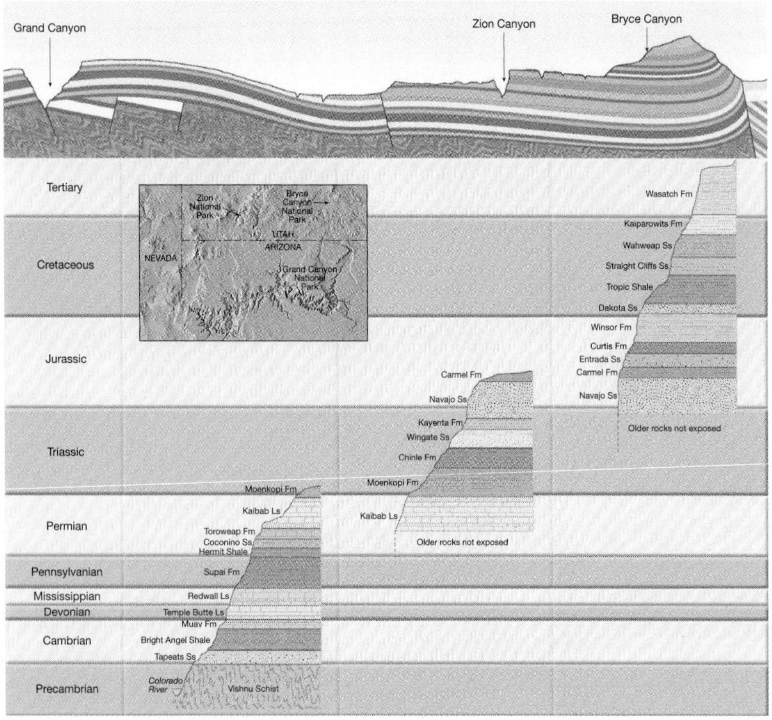

Figure 18.10 Correlation of strata at three locations on the Colorado Plateau.

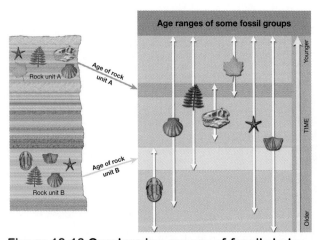

Figure 18.13 Overlapping ranges of fossils help date rocks.

NOTES:

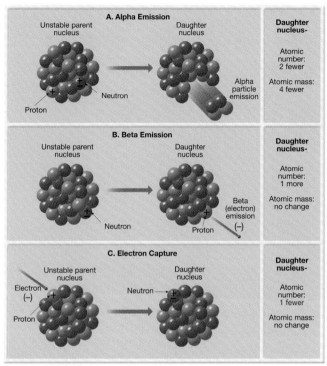

Figure 18.14 Common types of radioactive decay.

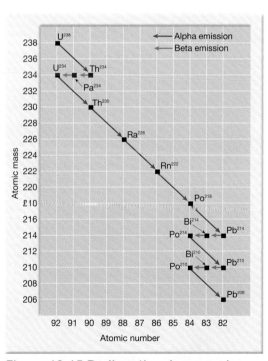

Figure 18.15 Radioactive decay series.

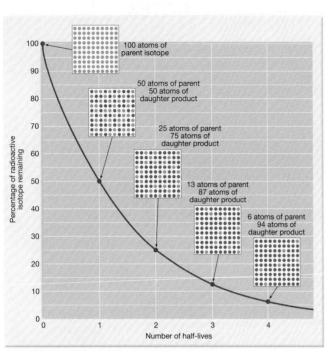

Figure 18.16 Radioactive decay curve illustrating exponential change.

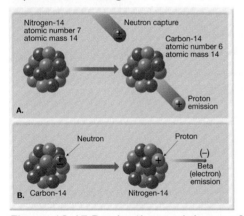

Figure 18.17 Production and decay of carbon-14.

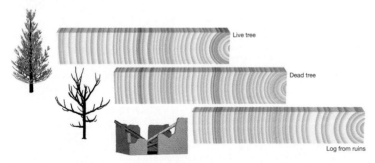

Figure 18.19 Cross dating is a basic principle in dendrochronology.

NOTES:

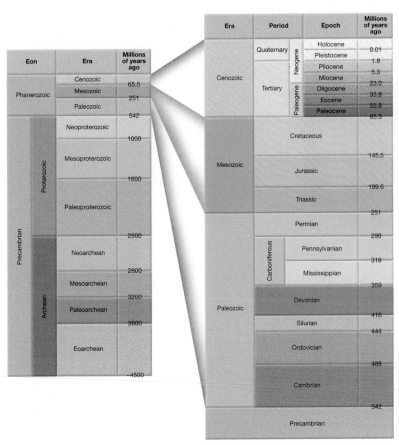

Figure 18.20 Geologic time scale.

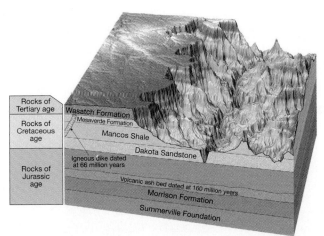

Figure 18.22 Numerical dates for sedimentary layers are usually determined by examining their relationship to igneous rocks.

NOTES:

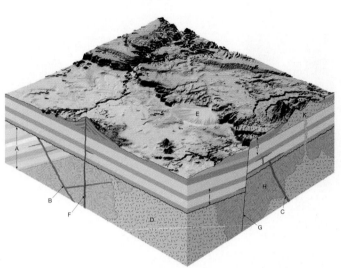

Figure 18.23 Hypothetical area in the American Southwest.

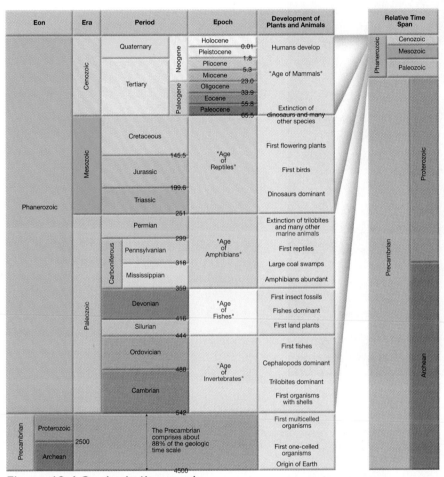

Figure 19.4 Geologic time scale.

NOTES:

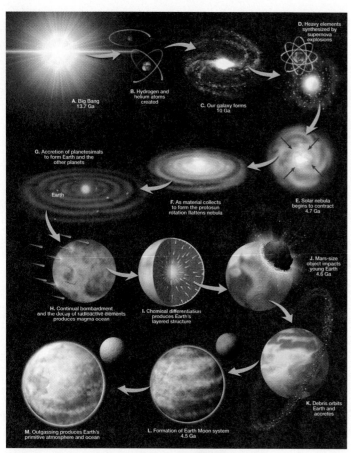

Figure 19.5 Formation of early Earth.

Figure 19.6 Depiction of Earth over 4 billion years ago.

NOTES:

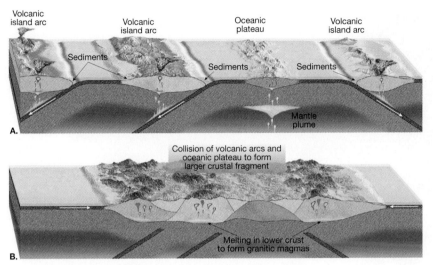

Figure 19.12 Collision and accretion of terranes.

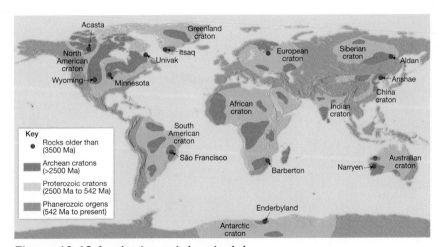

Figure 19.13 Ancient crustal material.

NOTES:

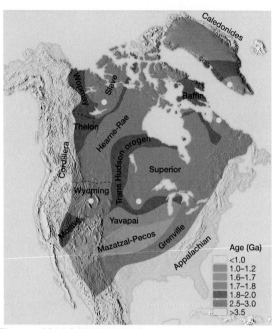

Figure 19.14 Major geological provinces of North America.

Figure 19.15 One of several possible configurations of Rodinia.

NOTES:

A. Continent of Gondwana

B. Continents not a part of Gondwana

Figure 19.16 Earth as it may have appeared in the late Precambrain.

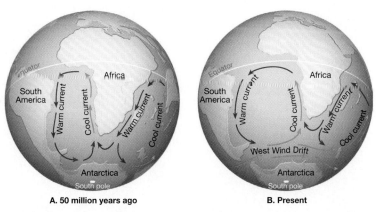

A. 50 million years ago   B. Present

Figure 19.17 Oceanic circulation: 50 million years ago and present.

NOTES:

A. Cambrian (500 Ma)

B. Silurian (425 Ma)

C. Mississippian (340 Ma)

D. Triassic (250 Ma)

Figure 19.19 Formation of Pangaea.

NOTES:

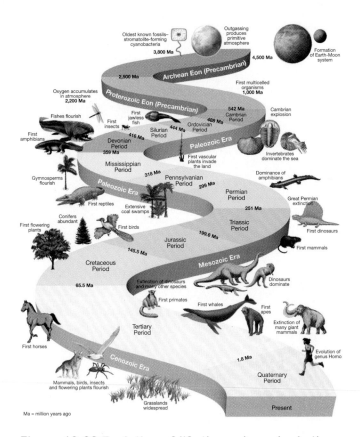

Figure 19.22 Evolution of life through geologic time.

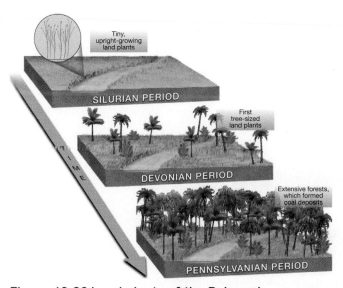

Figure 19.26 Land plants of the Paleozoic.

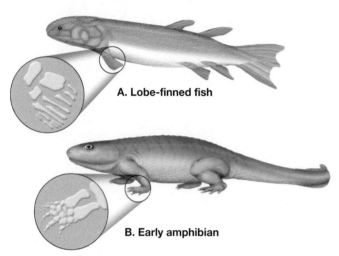

Figure 19.27 Lobe-finned fish and early amphibians.

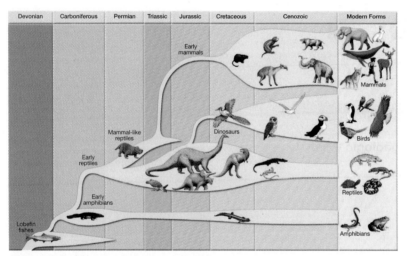

Figure 19.29 Vertebrate evolution.

Figure 19.30 Hypothesis for the "Great Permian Extinction."

NOTES:

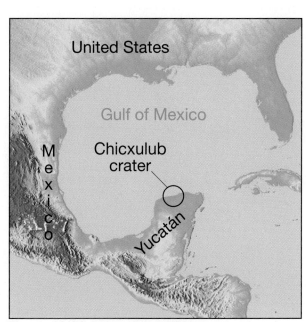

Figure 19.36 Chicxulub crater.

NOTES:

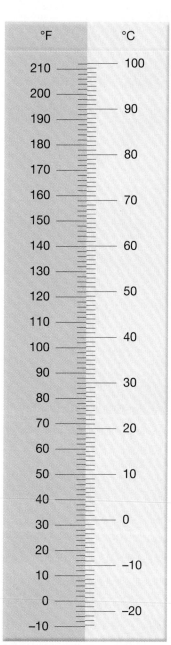

Figure A.1 Temperature scales.

Lutgens/Tarbuck, *Essentials of Geology*, 10e
© 2009 Pearson Prentice Hall, Inc.

NOTES:

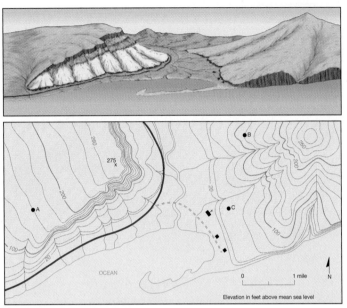

Figure B.1 Perspective view and contour map of an area.

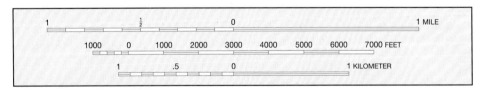

Figure B.2 Graphic scale.

NOTES:

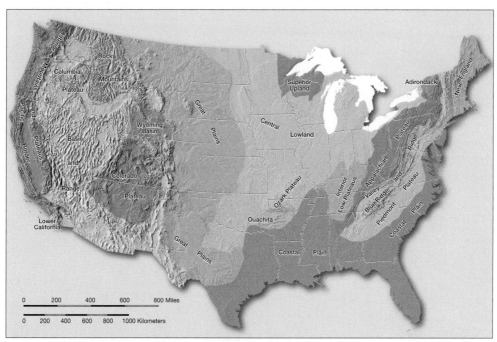

Figure C.1 Outline map of major physiographic provinces of the United States.

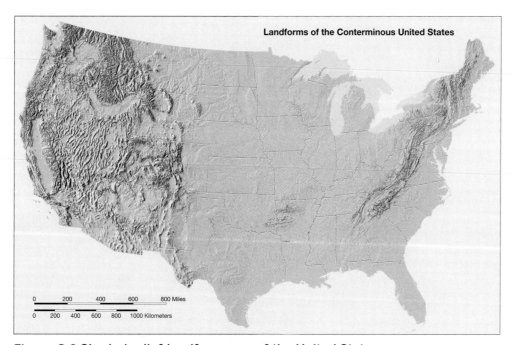

Figure C.2 Shaded relief landform map of the United States.

NOTES

NOTES

NOTES

NOTES

NOTES

NOTES

NOTES

NOTES